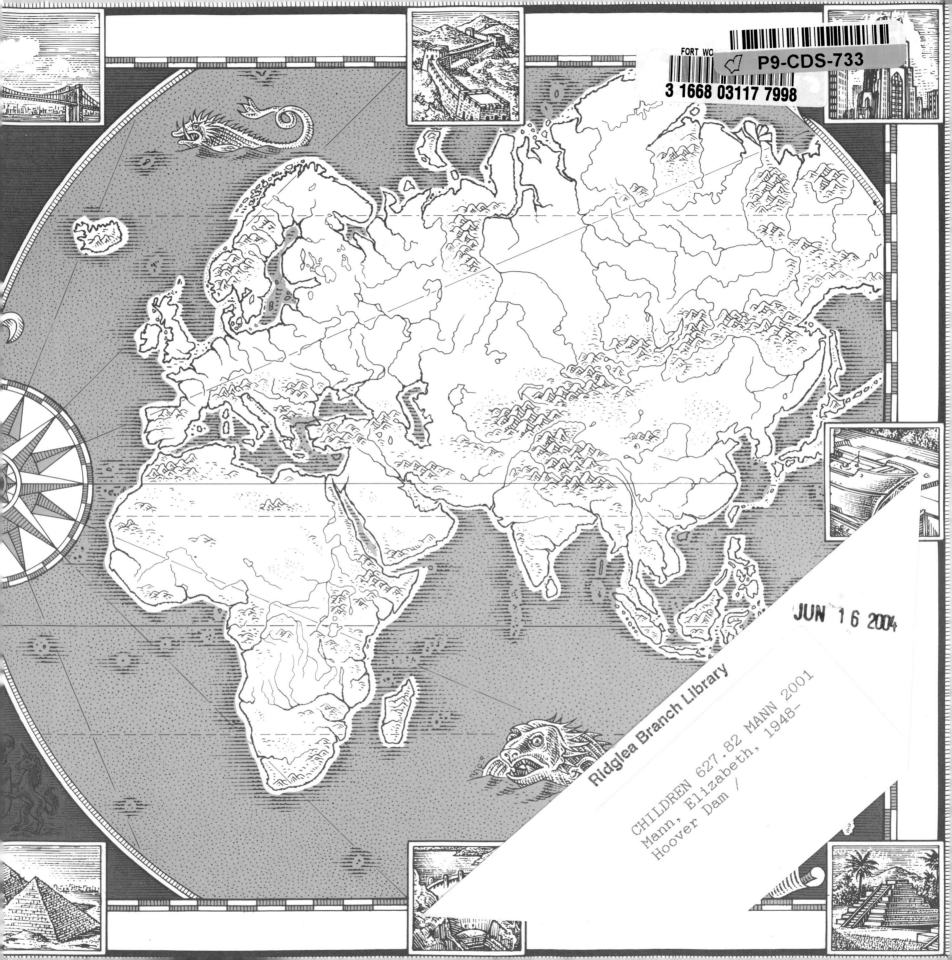

Dedicated to librarians everywhere who make the writing of nonfiction not only possible,
but a pleasure, and in particular to The Writers Room Librarian Alfred Lane.

Special thanks are due to Dennis McBride, Boulder City/Hoover Dam Museum Curator, researcher, oral historian extraordinaire,
for his extensive contributions to the creation of this book. The "Voices" of Hoover Dam and "Those Who Died" are taken
from his book (with Andrew Dunar), *Building Hoover Dam: An Oral History of the Great Depression.*

Thanks also to Kelly Conner of the Bureau of Reclamation in Boulder City for her invaluable assistance with photo research;
to Maureen Sullivan of the Boulder City/Hoover Dam Museum; to Joan Elizabeth Goodman; to University of Nevada Las Vegas Librarians
Peter Michel and Kathy War for help far beyond the call of good librarianship; to New York City Lab School 6th grade editors
Cindy Cheng, Magda Kelleher, Alanna Kowalski, Anny Oberlink and Catherine Rice for their insights and suggestions.

And thanks, as always, to The Writers Room in New York City for providing a quiet desk and a safe haven.

Other Books by Elizabeth Mann

The Brooklyn Bridge
The Great Pyramid
The Great Wall
The Roman Colosseum
The Panama Canal
Machu Picchu

Editor: Stuart Waldman
Design: Lesley Ehlers Design
Copyright © 2001 Mikaya Press
Illustrations Copyright © Alan Witschonke

Library of Congress Cataloging-in-Publication Data

Mann, Elizabeth, 1948-
    Hoover Dam/ by Elizabeth Mann ; with illustrations by Alan Witschonke.
      p. cm. — (Wonders of the World)
    Includes index.
    Summary: Describes the engineering, construction, and social and historical contexts of
the Hoover Dam.
    ISBN 1-931414-02-5
      1. Hoover Dam (Ariz. and Nev.)—History—Juvenile literature. [1. Hoover Dam (Ariz.
and Nev.)] 1. Witschonke, Alan, 1953-ill. II. Series.

TC557.5.H6 M35 2001
627'.82'0979313—dc21

                                                                                    2001034520

Hoover Dam

A WONDERS OF THE WORLD BOOK

By Elizabeth Mann

WITH ILLUSTRATIONS BY ALAN WITSCHONKE

MIKAYA PRESS

NEW YORK

It was a sure thing. Farmers would get rich if water could be brought to the Colorado River Desert.

The desert was a dry, desolate area covering 2,000 square miles of southern California. It had long been avoided by humans and animals, but it had deep, rich soil, a warm climate, and cloudless skies. The only thing standing in the way of it being an agricultural paradise was the lack of water, and that problem was solved in 1901 by an eager group of businessmen.

They dug a criss-crossing grid of irrigation ditches into the desert and linked them to the Colorado River. They spread the word that water was coming to the desert, and farmers rushed to California to buy land from them. A rough earth dam diverted the river from its natural course and the water surged into the ditches. As if by magic, the desert was transformed into lush farmland. Even the name was transformed—to Imperial Valley. Farmers harvested valuable crops of lettuce, grapes, and strawberries and sold them all over the country. Thanks to the Colorado River, they were indeed getting rich.

It was a happy time, but everything depended on the Colorado. No one wanted to hear warnings that the river was not a dependable ally, that it could flow quietly for years and then, without warning, flood violently. In 1904 it did just that.

Spring came early that year. High in the Rocky Mountains heavy rains fell, melting the snow and swelling thousands of mountain streams. The streams cascaded into rivers, the rivers roared into the Colorado, the Colorado overflowed its banks, and the flooding began. The earth dam was quickly washed away. The irrigation system, which had so efficiently delivered water and wealth to the Imperial Valley, now delivered destruction. Water filled the ditches and spilled over onto fields, washing away crops, eroding deep gouges through farms, destroying towns and homes, and tearing up roads and railroad tracks.

The flooding left behind ruined farms and towns and a strong determination to rebuild. Farmers had tasted the prosperity that an irrigated Imperial Valley could provide, and they weren't about to give it up. The Colorado had to be made useful.

The Imperial Valley disaster had shown that an earth dam would not work. Only a dam big enough and strong enough to block the river completely would allow humans to safely use the Colorado.

People have been building dams of all kinds and all sizes for thousands of years, and beavers have been building them for much longer. Whether they're made of sticks and mud or stone and concrete, all dams work in basically the same way: they block the flow of water and cause it to collect in a lake or pond.

A dam across the Colorado River would do exactly that and much, much more. From the beginning it was clear that Hoover Dam would be an extraordinary project. It would be, by far, the largest dam in the world. It would also be the first ever multipurpose dam, protecting against flooding, providing irrigation water, and producing electricity.

Rivers naturally flood when rainfall is heavy and snowmelt is high. It's only a problem when people live and farm in the areas where flooding occurs.

A dam blocks a river and creates a lake behind it. Water stored in the lake can be released in safe amounts so that flooding is no longer a danger. Irrigation ditches can be filled whenever crops need water, so that farmers don't have to depend on rainfall for successful crops.

Building Hoover Dam would require taming one of the most dangerous and unpredictable rivers in the world. The Colorado had never been friendly to humans. The river's fast current and swirling rapids made boat travel impossible along much of its 1,400-mile length. Many of its canyons were virtually inaccessible to any creature less nimble than a mountain goat. The river contained so much silt that it took hours for the sediment to settle to the bottom of a glass of its water, and even then it still had a muddy taste and the reddish-brown color that gave the Colorado its nickname—the Red Bull.

The Colorado flowed through countryside that was just as unfriendly. The Southwest is the driest region of the United States. There's barely enough rainfall to grow grass for grazing cattle, and farming is out of the question. In the summertime the heat is stifling. There were no roads, railroads, electricity, or towns in the area—everything would have to be built from scratch. Hoover Dam presented a real challenge to the engineers who would design it and the workers who would build it.

The Bureau of Reclamation, a department of the United States government, was responsible for "reclaiming" arid land. It created farmland by damming rivers and using the water for irrigation. Bureau engineers had built dozens of dams, but they had never built one so large or faced such a dangerous river. It was an enormous undertaking, and the eyes of the world would be on them. They proceeded carefully.

The deepest of all the Colorado's canyons, Grand Canyon, is less than 100 miles upstream from Hoover Dam.

The first step was to find the best possible location—one where the greatest amount of water could be safely stored with the least effort and expense. In the right location, a tall, narrow dam could contain as much water as a tall, wide dam, and it would be faster and cheaper to build. A deep, narrow canyon was needed. Thanks to millions of years of erosion by the Colorado River, there were many to choose from. Wilderness exploration began in 1920, and the search for the best canyon eventually narrowed down to two: Boulder Canyon and Black Canyon.

The Bureau of Reclamation sent crews of surveyors and mapmakers to study possible sites in both canyons. The work was hard and dangerous. Surveyors dangled from ropes a thousand feet above the muddy current, taking measurement after measurement on the cliffs. Drillers struggled to keep their footing on wave-tossed rafts as they drilled rock samples from dozens of different spots in the river bed. At night the wind ripped the tents where the men slept. During the day it was so hot that tools burned their hands and thermometers shattered. The work could also be deadly. On December 20, 1922, driller J. G. Tierney drowned when the churning river swept him from his raft.

After years of study, the Bureau selected a site in granite-walled Boulder Canyon. Plans were already being drawn for Boulder Dam when unexpected weaknesses, called faults, were discovered in the granite. The great weight of a concrete dam on these faults could have caused the rock to shift. It was too much of a risk to build the dam on rock that might be unstable. The Bureau had no choice but to look for another location.

A site in Black Canyon, 20 miles downriver, was chosen instead. There the bedrock and canyon walls were of andesite breccia, another volcanic stone, but one that was solid and free of faults, strong enough to support the weight of the 6,600,000-ton dam forever. (Confusion about the dam's name seems destined to last forever also. Even though the location was changed to Black Canyon, and the name was later changed to Hoover Dam, some people still call it Boulder Dam!)

In Washington, D.C., far from the Colorado's wild canyons, men in dark suits and crisp, white shirts gathered in meeting rooms to deal with other, very different problems that had to be solved before the dam could be built.

The Colorado flowed through seven southwestern states on its way to the ocean. Each one had the right to use water taken from the river. Water was precious in that arid land, so naturally each state wanted as much as possible. How would the water be shared? The arguments were fierce and went on for years.

Electricity was becoming as important as water in the growing southwest, and everyone wanted as much as they could get. The power generated at Hoover Dam would be sold to pay for its construction. Who would be allowed to buy it? How much would they pay? More arguments.

It wasn't easy, but by the end of 1930 many of the problems had been solved. The seven Colorado Basin states had reached an understanding about the water, and the sale of electricity had been arranged. Design engineers at the Bureau of Reclamation were hard at work drawing plans for the dam in Black Canyon, and construction had begun on a railroad to bring supplies to the site.

One very important decision still had to be made. Who would build the dam? Dozens of construction companies were interested at first, but the job was so large and complicated that most were quickly frightened off. In the end, only a handful came forward. One of them, Six Companies, was actually a group of smaller companies that had joined together just to try to win this job. Their bid, the amount they would charge the Bureau of Reclamation to build the dam, was the lowest. Six Companies won the contract to build Hoover Dam.

Many rivers, large and small, feed into the Colorado. The area from which all of them collect rainfall and snowmelt is the drainage basin of the Colorado River. The basin is enormous—one-twelfth the area of the United States.

While preparations for the dam were going on, the United States was being rocked by a thunderous change. In October of 1929, the New York Stock Exchange "crashed." Suddenly businesses that had been worth millions of dollars were worthless. People who had invested their life savings in those companies were now penniless. People who had worked for them were unemployed. It was the beginning of the Great Depression.

During a depression, jobless people have no money to buy things, so more companies go bankrupt, and more people are thrown out of work. The companies that manage to stay in business hire fewer people and at lower salaries. It's a hard time for everyone.

There had been other economic depressions in the United States, but the Great Depression was by far the worst. One quarter of all American workers were unemployed. Farmers lost their farms, banks closed, and families' life savings disappeared.

Soup kitchens were sometimes the only source of food. Children waited in line with containers—in this case empty paint cans—to receive their allotment of soup.

People who were evicted from their homes often had no choice but to live on the street.

Some people who couldn't afford cars traveled by train, riding in empty freight cars. "Hopping a freight" was dangerous, illegal, and uncomfortable, but it was free.

Millions of people were suddenly poor, homeless, and desperate. They piled mattresses and cooking pots onto the roofs of their cars and roamed the country looking for work.

Rumors about a gigantic dam, and lots of construction jobs, spread quickly. Model T Fords from all over the country headed for southern Nevada as early as the summer of 1930, long before work on the dam was due to begin.

Some job seekers ended up in Las Vegas, Nevada, 30 miles from Black Canyon. Las Vegas was a small town then, only 5,000 residents, and there was little room for the thousands of newcomers. People slept in their cars and on the grass in front of the courthouse.

Some moved closer to the dam site, creating a camp called Ragtown on the banks of the Colorado. They made tents from blankets and built shacks from scraps of wood. They used the muddy river water as best they could for drinking and bathing. They tried to stay cool in the 100° heat by sitting in the river or lying in the only shade available—underneath their cars. And they waited.

On March 11, 1931, a tall, thin man stepped from a train as it arrived in Las Vegas station. A broad-brimmed hat shielded his eyes from the desert glare. Frank Crowe was an engineer and the most brilliant dam builder in the country. He had been hired by Six Companies to be the superintendent of construction on Hoover Dam. He was the key to the success of the project. Without him Six Companies probably would not have won the contract for the dam, much less been able to build it. Crowe had already built six dams in his career, but this was the one he had been dreaming of for years.

To the men who waited, Crowe's arrival was a reason for hope. It meant that the rumors would end and the hiring would begin. As he approached the Six Companies temporary office in Las Vegas, Crowe had to make his way through the groups of unemployed who crowded around the door. The next day in Ragtown, a similar crowd gathered around him. Hungry men begging for work—that was something Crowe had never encountered before.

Herbert Hoover was also interested in jobs at Hoover Dam. As president of the United States, he was being held responsible for the Great Depression and for the hardship that it was causing. He was stung by the accusations and felt the pressure to ease the suffering. Hoover Dam was an opportunity to put thousands of people to work, and he jumped at it. Preparations were not finished, and construction was not scheduled to start until October, but he gave the order to begin work immediately. His order provided jobs quickly, as it was intended to do, but at great cost to the workers.

VOICES

Everyone who worked down here knew who Frank Crowe was. He was all over the job. His workmen, he knew them by their first name, nearly every one of them. . . . Once he learned your name he never forgot it. Any time you saw a white shirt and a large Stetson hat coming, you knew that was Frank Crowe. Very tall and erect, kind of a stately looking person, a very likable fellow.

BOB PARKER
Dining Hall Worker

There was no place for them to live, no pure water to drink, no hospital to treat their injuries. Although Six Companies was required by their contract with the Bureau of Reclamation to provide all these things, it would be many months before everything was ready. Until then workers would be living and working in unsafe, even life-threatening conditions.

Frank Crowe followed orders and rushed ahead with the dam. He truly earned his nickname, "Hurry Up Crowe," that spring. By the middle of May, he had 1,100 men working day and night.

Down in Black Canyon, Frank Crowe and his crews were contending with the mighty Colorado. The dam could not be built in the middle of a fast-flowing river. The Red Bull had to be diverted away from the construction site. The problem was—where could the diverted water go?

The only way to carry the Colorado away from the site of the future dam was in tunnels blasted through the solid rock walls of Black Canyon. Four enormous tunnels would be needed, two on the Nevada side of the river and two on the Arizona side.

Once the tunnels were ready, earth and rock would be dumped into the Colorado, making a temporary dam called a cofferdam. When the water rose behind the cofferdam it would flow into the upper portals, through the tunnels, and back out into the riverbed through the lower portals. Another cofferdam below the dam site would prevent any water from splashing back into the construction area. Between the two cofferdams would be a dry worksite half a mile long.

Upper portals

Upper cofferdam

Future dam

NEVADA

Diversion tunnels

Diversion tunnels

ARIZONA

Lower cofferdam

Lower portals

Lower portals

In May 1931, eight white circles 56 feet in diameter were painted on the dark canyon walls. Like giant bull's eyes they marked the upper and lower portals of the four tunnels. Work began at both ends of each tunnel. Drillers bored holes deep into the white circles. Dynamite was packed into the holes and exploded, shattering the solid rock into rubble. "Muckers" moved in with shovels, bulldozers, and trucks to clear away the rubble. Then the drillers returned to drill another set of holes, and so the tunnels pierced deeper into the rock.

Digging the tunnels and diverting the river through them was a big job. Bureau of Reclamation plans called for it to be finished by October 1, 1933. After that Six Companies would be charged a penalty of $3,000 for every day they were late. With that much money at stake, "Hurry Up Crowe" hurried, and when he did, everyone who worked for him did, too.

Speed was often valued more than safety. Carbon monoxide gas from truck engines lingered in the poorly ventilated tunnels. Clearing the fumes from the tunnels would have wasted valuable time, so workers were expected to work alongside the trucks. Many were poisoned by the deadly air. Tunnel temperatures rarely went below 130° and men collapsed from the heat. The seven-day work week was exhausting. The nights were too hot for sleep, so the men couldn't regain their strength after a punishing day's work. Because there hadn't been time to build a hospital, some died for want of basic medical care. By the end of July, at least 20 workers had died, and many others had become ill or been injured.

Women and children who stayed behind in the shadeless squalor of Ragtown were suffering also. The heat and bad drinking water were taking a toll on them, and deaths began to mount. Danger on the job and misery at home were pushing workers toward a breaking point.

That breaking point was reached in August of 1931 when Six Companies announced that the muckers, already the lowest paid workers on the job, would have to take a cut in pay. Angry workers walked off the job and refused to return until their demands had been met. For example, they wanted Six Companies to obey the Arizona and Nevada mining safety laws, which prohibited the use of carbon monoxide-producing trucks in tunnels. They wanted cold drinking water to be available on the job. Nowadays such demands would seem like common sense, but during the Depression workers' health and safety were often ignored. People were desperate for any job at all, no matter how dangerous, and they weren't likely to complain.

Many of the men in Black Canyon had worked with Frank Crowe before. His "construction stiffs," as they called themselves, had followed him from dam to dam around the country. Crowe had always been a fair boss, and they expected that he would speak to the owners of Six Companies on their behalf. They hadn't realized how important Hoover Dam was to him. To their surprise he fired everyone, including the loyal "stiffs," and closed down the job.

Without Crowe's support, and with thousands of unemployed workers in Las Vegas eager to take their places, the strikers didn't stand a chance. After just a week, most returned to the Six Companies payroll, and to a workplace that was as hazardous as when they left. Speed had won out over safety. In the end only one change was made—Six Companies began providing drinking water on the job.

VOICES

...four women died in one day. That was the 26th of July [1931] and it was terribly hot.... there was a woman that was 28 that was just three tents from me.... she was just lying across a folding cot and she was dead.... There were three of us that went over into her tent. We kind of straightened her up on the bed. I went back to my tent and I told my husband, "I've got to get somewhere I can get the babies to a doctor if need be, and also myself.

ERMA GODBEY
Ragtown Resident

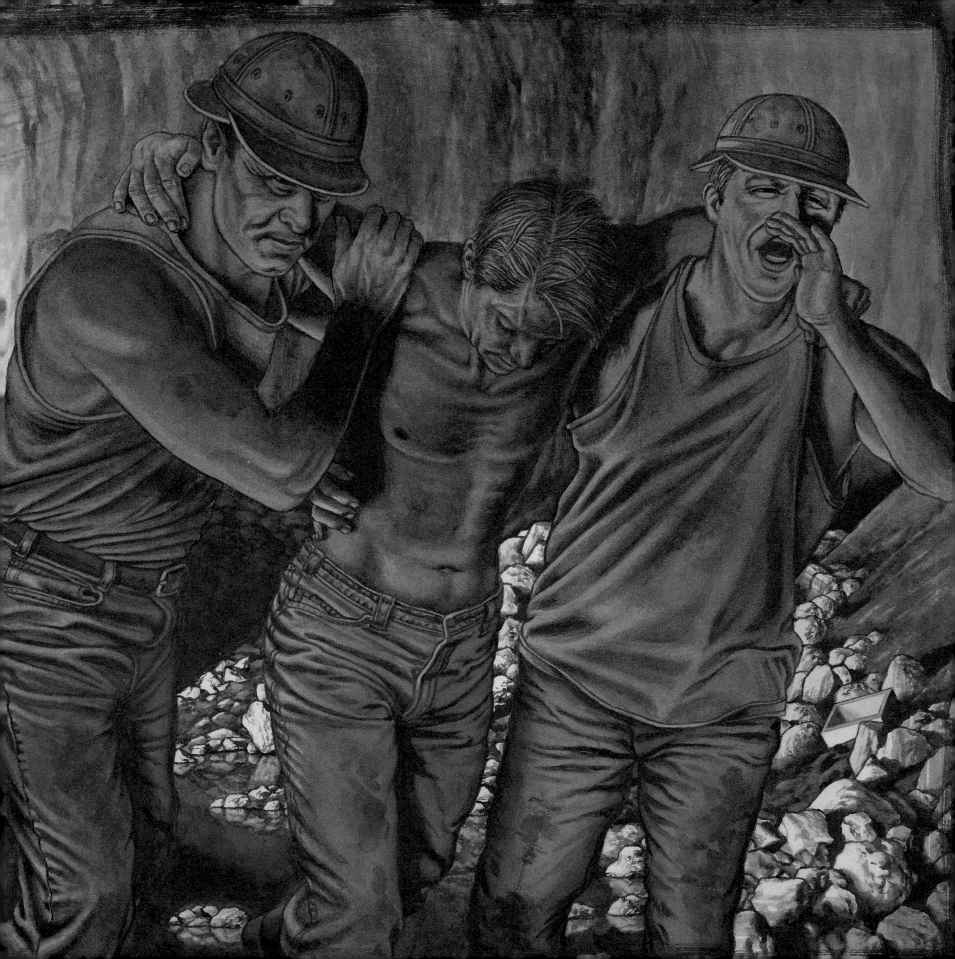

High above the river, work of a different kind was going on. Giant grooves were being cut into the canyon walls, notches that would securely hold each end of the concrete dam. There were also areas of loose rock—the result of years of weathering—that had to be removed. Even a small piece of rock falling from a great height could crush the skull of a worker below. Drilling and blasting the rock on the canyon walls was the job of the "high scalers."

High scalers dangled like spiders hundreds of feet above the river on little wooden swings tied to the end of a single rope. Their waist belts were heavy with tools and water bags. They lowered themselves down the cliff face, turned on their noisy jackhammers and went to work drilling the rock and placing dynamite charges. Thousands of tourists who visited the dam site every year were captivated by the daring high scalers, and the high scalers delighted in an audience. When their supervisors weren't watching, they did tricks on their swings that had the crowds gasping and applauding.

Although the possibility of a plunge to certain death was always present, falling rocks and tools proved to be a greater danger, and workers devised a way to protect themselves. They placed one cloth hat (like today's baseball cap) inside another, so that the brims faced in opposite directions. Then they soaked the doubled hat in tar and let it dry, creating hard hats. They worked so well that Frank Crowe ordered factory-made hard hats for everyone on the job.

VOICES

But that was a good job. I got paid $5 a day to start with. Afterwards I got $5.60. I believe that was one of the safest jobs they had. I think there was less people got hurt on high scaling than there was on lots of other jobs. It wasn't worse than anything else. One thing, you were sitting down all the time. It was a sitting down job.

JOE KINE
High Scaler

Eight miles from the dam, another big job was under way. From the workers' point of view, it may have been the most important. There on a hillside where wild burros and tarantulas roamed, a town was being built for them. The location, in the middle of nowhere, was parched and barren, but a steady breeze kept it around ten degrees cooler than Black Canyon.

Boulder City was a unique experiment: a community designed, built, and run by the U.S. government for 5,000 workers, many with families. Everything, every house and tree and park bench, was planned out ahead of time. Just as Hoover Dam was a hopeful sign that once again there would be jobs for the jobless, so Boulder City was a sign that there would also be homes for the homeless. The town's progress was watched with interest around the country.

Sturdy brick houses were built for Bureau of Reclamation employees who would stay in Boulder City to maintain the dam after it was finished. The Six Companies homes were flimsier. The plan was to tear them down once the dam was finished and the workers moved on.

By 1932 Boulder City began to resemble a real town. It had homes, dormitories, running water, a sewer system, a police station, a hospital, a dining hall for dam workers, offices, a town hall, and, on the highest point,

VOICES

These houses were put up in one day, child. You think you can put a house up in one day and have it look like anything? They were never comfortable because there was no insulation in 'em. And the sheetrock inside was so thin—you sneeze on it, you'd blow a hole in it. But they were never meant to stay here. It was all supposed to revert back to desert when the dam was finished.

WILMA COOPER
Boulder City Resident

a sprawling home for Frank Crowe and his family.

Boulder City rules were strict and sometimes puzzling. Children had to go indoors when the curfew siren blew at 9 p.m. Wives were not allowed to hold jobs. People who failed to water their lawns were charged extra for their drinking water. The hospital treated dam workers, but not their wives and children. As a result many babies were born on the tiny sleeping porches of the Six Companies houses.

The impact of one rule was felt by people who didn't even live there: Only white Americans were allowed in Boulder City. A daily commute on the rough dirt road between Las Vegas and Black Canyon was so difficult that the rule also served to keep nonwhites from working on the dam. Although segregation was still a fact of life in the American south, many people were critical of discrimination on a U.S. government project. President Franklin D. Roosevelt, who had replaced Herbert Hoover, was among them. Six Companies responded to the criticism and reluctantly hired a handful of black laborers. They were assigned to the least desirable jobs and were never permitted to set foot in Boulder City.

For the whites who were allowed to live there, the benefits of Boulder City were immediate. In the relative coolness and comfort of the town, workers were able to recover from their grueling shifts in Black Canyon. There were no more deaths from heat exhaustion.

Every house was exactly alike. You couldn't tell your own house. It was always a joke in the olden days about somebody coming into the wrong house. I do know of cases where people got up in the morning and found a man sleeping on their couch. But they'd just wake him and ask what he was doing. "My gosh, this is not where I belong!"

ROSE LAWSON
Boulder City Resident

Boulder City rules were written and enforced by Sims Ely, a Bureau of Reclamation employee whose job it was to run the town. There was no elected government. Ely was sheriff, mayor, judge, and jury all rolled into one. He was allowed to evict families from Boulder City for minor offenses and often did. Since being thrown out of town meant being thrown out of work, his power was widely feared.

Houses could be built in a day, but lawns took longer to grow. Children got used to playing among the cactuses in their sandy front yards.

Early in the morning, day shift workers climbed aboard the specially built, 150-passenger trucks, Big Berthas, that carried them from Boulder City to the dam. There they replaced the "graveyard shift" workers, who rode the Big Berthas back to Boulder City, put a "Day Sleeper" sign on the front door, and collapsed into bed. Eight hours later the trucks carried the "swing shift" to the job and picked up the exhausted day shift. Christmas and the Fourth of July were the only days off.

 VOICES

Then, of course, we had what we called the Donkey Baseball Team. They'd hit the ball, then climb on the burro and try to turn around first, second, and third base without getting knocked off or pushed off or the donkey falling over.

SAUL "RED" WIXSON
Steam Shovel Operator

On January 29, 1932, a round of dynamite charges exploded, as usual, in one of the tunnels. The noise from the blast echoed along the walls and gradually faded. As the muckers moved in to clear away the broken rock, they saw lights through the gritty dust and felt a rush of air. They had broken through! Breakthroughs in the other three tunnels followed quickly.

The river couldn't be diverted directly into the bare stone tunnels. Billions of gallons of rushing, silt-heavy water would have eventually found weaknesses even in the hard andesite breccia of Black Canyon. Constant scouring by the Red Bull could have caused weakened sections of tunnel to collapse. To prevent this each tunnel was lined with a layer of concrete three feet thick.

In November of 1932, all four tunnels were ready. After a dry summer, the river was at its low autumn level, as meek and unthreatening as it would ever be. A seemingly endless caravan of trucks dumped boulders, rocks, and dirt—rubble that the muckers had removed from the tunnel—into the Colorado just below the upstream portals. Load after load splashed into the muddy water, one every 15 seconds, for an entire day and night. Slowly the cofferdam rose through the water and blocked the river. At last, on November 14, it poured obediently into the waiting tunnels.

Success! The Colorado was turned from its ancient course and diverted around the dam site. With the river roaring safely through the canyon walls, Crowe and his crews could breathe easier. The threat of a flood had hung ominously over Black Canyon during the tunneling. At any time the unpredictable Colorado could have risen up and destroyed equipment, taken lives, and delayed the job indefinitely. They had worked hard, but they had also been very lucky—the Red Bull had cooperated. The river was diverted a year ahead of schedule. For Six Companies the hurrying had paid off.

The newly dry dam site was still not ready for construction to begin. Like any large structure, Hoover Dam had to be built on a solid foundation of bedrock. Bedrock in Black Canyon lay buried beneath 135 feet of silt and stones that the Colorado had deposited over millions of years. The muckers who would dig it out had eight months of hard labor ahead of them.

June 6, 1933, was the big day. The surface of the exposed bedrock had been swept and sponged until it was free of any traces of mud. Wooden forms, like rooms without ceilings or floors, were waiting in place on the bedrock at the very bottom of the dam site. The forms would hold the concrete in place until it hardened. The two long years of preparation were over. It was time to actually build the dam.

Officials from Six Companies and the Bureau of Reclamation peered skyward as an enormous bucket of concrete swung gracefully through the air suspended from a steel cableway that stretched across Black Canyon. Photographers clicked their shutters as the bucket was slowly lowered an eighth of a mile to the bottom of the canyon.

"Puddlers," workers in high rubber boots, waited in an empty form until it hovered in position next to them. Then, at a signal from Frank Crowe, they released the safety latches on the bucket's trap doors and jumped out of the way. Sixteen tons of wet concrete dropped in an abrupt glob into the form.

Relieved of the weight, the bucket bounced skyward on its cable like a yo-yo. The puddlers leaped into action just as quickly with shovels and spreaders, stomping and smoothing the concrete, making sure there were no air bubbles to weaken it after it hardened.

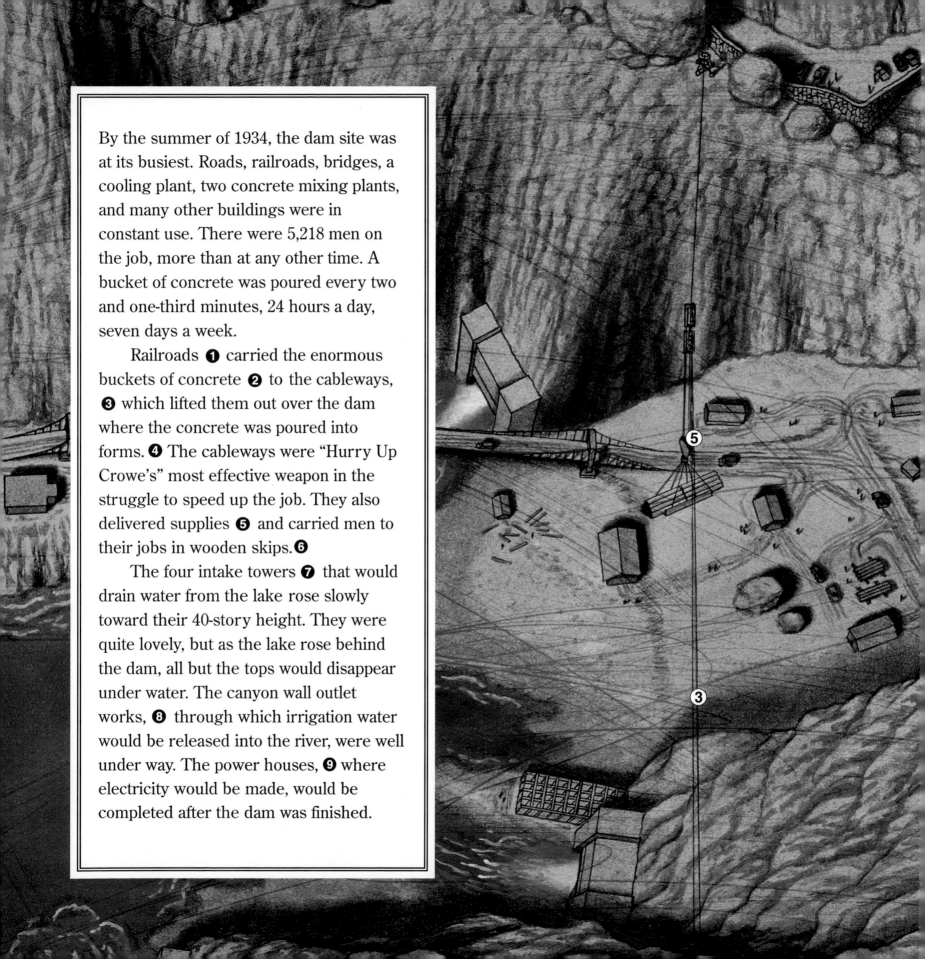

By the summer of 1934, the dam site was at its busiest. Roads, railroads, bridges, a cooling plant, two concrete mixing plants, and many other buildings were in constant use. There were 5,218 men on the job, more than at any other time. A bucket of concrete was poured every two and one-third minutes, 24 hours a day, seven days a week.

Railroads ❶ carried the enormous buckets of concrete ❷ to the cableways, ❸ which lifted them out over the dam where the concrete was poured into forms. ❹ The cableways were "Hurry Up Crowe's" most effective weapon in the struggle to speed up the job. They also delivered supplies ❺ and carried men to their jobs in wooden skips.❻

The four intake towers ❼ that would drain water from the lake rose slowly toward their 40-story height. They were quite lovely, but as the lake rose behind the dam, all but the tops would disappear under water. The canyon wall outlet works, ❽ through which irrigation water would be released into the river, were well under way. The power houses, ❾ where electricity would be made, would be completed after the dam was finished.

The slow drama of that first pour marked the beginning of 30 months of frenzied activity. As the workers became more skilled at mixing, transporting, pouring, and puddling the concrete, the pace picked up. Concrete had never been poured so fast, and Frank Crowe made sure the job went even faster. He offered bonuses to the crews that poured the most concrete in one eight-hour shift. More speed meant more accidents. The cableways were in constant motion, whisking supplies, men, and buckets of concrete through the air. A moment's carelessness or confusion, a worn cable, or a missed signal could result in injury or death.

Such a large mass of concrete had never been poured in one place at one time, and it presented a problem that had never been faced before.

Concrete is a mixture of cement, sand, water, and aggregate (small rocks). When the mixture hardens, it doesn't simply dry out, as mud and clay do. Instead, a chemical reaction takes place between the water and the limestone in the cement. As the chemical reaction occurs, it gives off heat. In most structures the concrete cools quickly but, because Hoover Dam was so massive, it was estimated that it would take 125 years to cool completely. As it cooled slowly and unevenly, the concrete would have cracked, weakening the dam. This had to be prevented.

An ingenious cooling method was devised for the dam. Workers laid a network of one-inch copper pipe in the forms before each bucket of concrete was poured. Icy water was pumped continuously through the pipes, turning each form into a kind of refrigerator. When the concrete was cool, the water was drained and the pipes were filled with grout, a thin liquid concrete. Over 528 miles of grout-filled pipe remain sealed inside the dam.

The dam was made up of many individual concrete columns. Once the concrete in a form had hardened, the form was taken apart and rebuilt on top of the hard concrete. More concrete was poured, and foot by foot, the columns grew taller. The spaces between the columns were filled with grout, making the dam a single, solid structure.

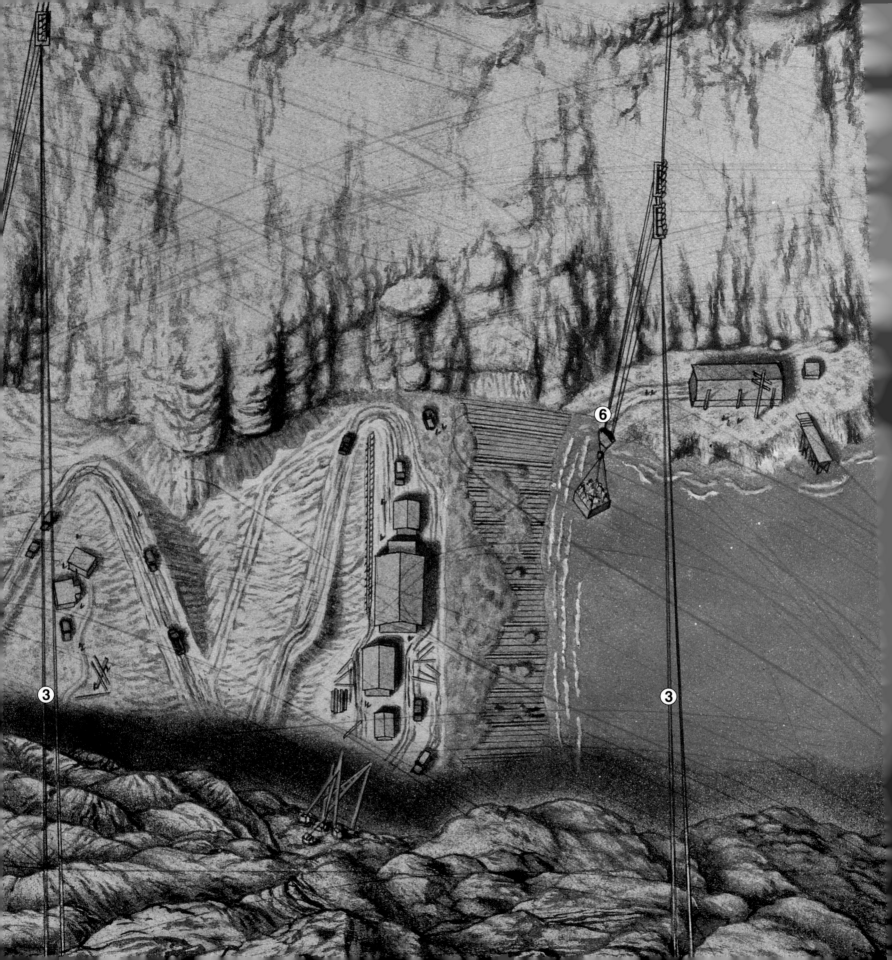

The spillways were enormous, expensive, and took months to build. They have only been needed once (above) during a flood in 1983 that lasted for two months.

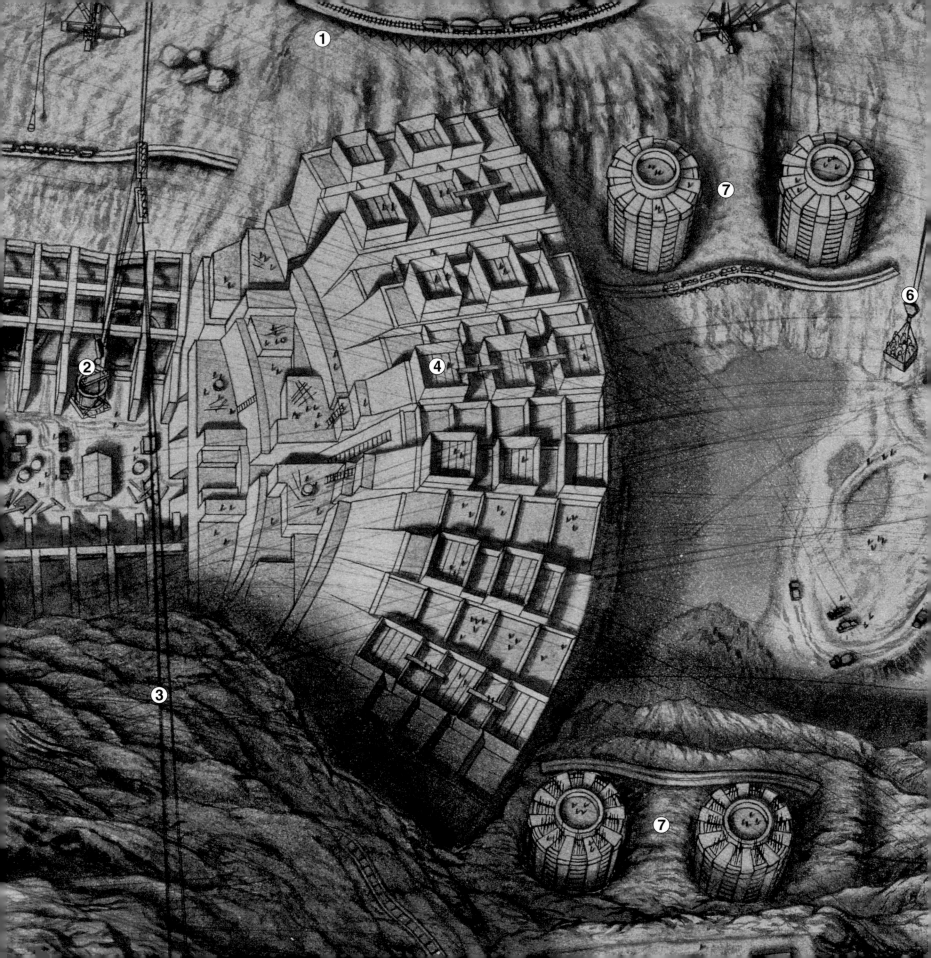

The Godbeys, a Boulder City family, look like miniature dolls inside this 170-ton penstock section.

Two emergency spillways were built alongside the lake. They were a mark of the respect that the mighty Colorado still inspired. Their sole purpose was to prevent flooding in case the water level in the lake ever rose high enough to spill over the top of the dam. Water pouring over the face of the dam would have quickly destroyed it, causing disastrous flooding all the way to the Imperial Valley and beyond. If the level of the lake ever came within five feet of the top of the dam, the water would drain into the spillways, just as water in a bathroom sink flows into the overflow drain before it can spill over onto the floor. The water would be carried safely past the dam through two of the diversion tunnels.

Deep inside the canyon walls, more tunnels were blasted through the rock to carry lake water from the intake towers to the power houses below the dam. Steel pipes, called penstocks, were placed inside all the tunnels. The penstock size varied depending on the size of the tunnel. The largest ones were 30 feet in diameter, bigger than any that had ever been made before. Because they were too big to be carried by train to Black Canyon, a factory was built to manufacture them on the spot.

By February 1, 1935, the dam was large enough to begin doing the job it was intended to do. One by one the diversion tunnels were plugged. Slowly, almost imperceptibly, the water rose behind the dam and a new lake, Lake Mead, began to form.

Once the lake had risen high enough and the first of 17 generators were installed in the power houses, the dam could begin producing electricity.

Water from the lake would flow from the intake towers ❶ into the penstocks inside the tunnels. ❷

As it raced downhill through the penstocks, the water would have tremendous power, in the form of mechanical energy. The mechanical energy of rushing water is not easy for people to use, but in the power houses, ❸ the mechanical energy would be turned into a form of energy that is very useful—electrical energy.

The rushing water would spin turbines. ❹ The turbines would spin generators ❺ and the generators would produce electricity.

Wires ❻ would carry the power of the Red Bull, in the form of electricity, to cities hundreds of miles away to run everything from light bulbs to factories.

If water was needed for irrigation but not for electricity, different penstocks would carry it past the power houses and release it directly into the river. Some would be released through the canyon wall outlet works ❼ and some through the lower portals ❽ of tunnels ❾ that had originally been used to divert the Colorado. The water would continue downstream to farms in the Imperial Valley. If ever the lake rose dangerously high, the water would flow into the spillways, ❿ through the spillway tunnels ⓫ (also originally used for diversion), and back into the river through the the lower portals. ⓬

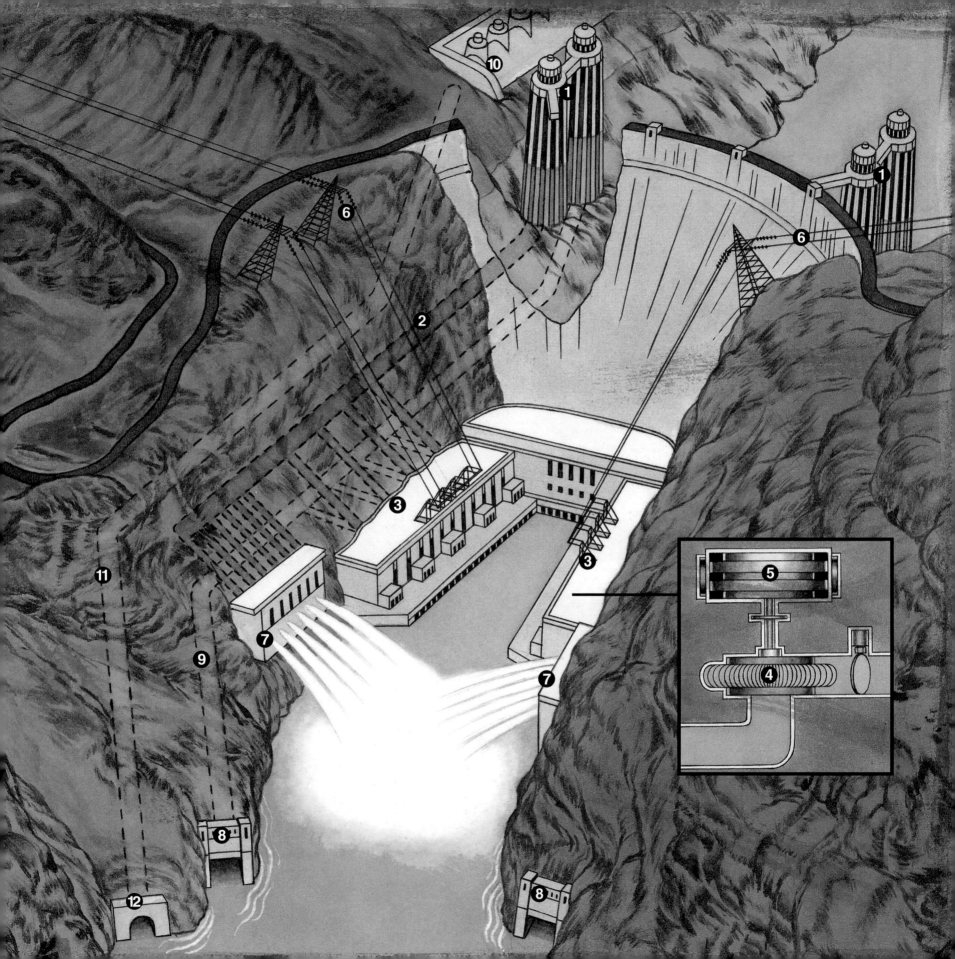

By early 1935 there was less activity on the dam site. The graveyard shift was eliminated. Boulder City began to empty out as workers and their families looked elsewhere for employment.

Frank Crowe was among the people who moved on as the work slowed down. He had done his part on Hoover Dam and he was looking for new challenges, new dams. He may have suspected that he would never again build anything as important as Hoover Dam. He was not an old man, but he had already accomplished the greatest work of his life.

Black Canyon was feeling the effects of changes taking place in the rest of the country. The cruel pinch of the Great Depression was easing, and there were more jobs. The threat of being fired wasn't as frightening as it had been in 1931, and workers were no longer willing to risk their lives for their paychecks. The furious pace on the job was relaxed. Safety, at last, was winning out over speed.

On September 3, 1935, President Roosevelt arrived in Black Canyon to officially dedicate the dam. A shiver of pride ran through the crowd of 20,000 gathered for the ceremony when Roosevelt said, "This is an engineering victory of the first order—another great achievement of American resourcefulness, skill and determination." Many who listened were dam workers who knew only too well the struggle and sacrifice that had gone into Hoover Dam. The victory was truly theirs to claim.

Twenty million people across the country listened to the president's speech on the radio, and they too shared in the victory. Hoover Dam had been a symbol of hope during the worst years of the Great Depression. Its dedication meant that the hard times were really over.

The dam was dedicated, but work was still being done. Four months after Roosevelt made his speech, on December 20, 1935, an electrician named Patrick W. Tierney fell to his death from an intake tower. He was the last person to die on the Hoover Dam project. By a stranger-than-fiction coincidence, his father, driller J. G. Tierney, who drowned exactly 13 years earlier on December 20, 1922, had been the first.

Today the massive white curve of Hoover Dam gleams dizzily in the desert sun. An orderly line of traffic winds down the highway into Black Canyon and across the top of the dam. Behind it Lake Mead, a brilliant blue now that the Colorado's silt has settled to the bottom, stretches for 115 miles. Hundreds of feet below its still surface, the remains of Ragtown lie undisturbed. Hundreds of miles downstream, Imperial Valley crops ripen in the sun. Inside the power houses, generators hum evenly and workers move silently across the marble floors. The clamor and pandemonium of the construction years, the 112 deaths, the thousands of accidents, all seem a distant and impossible memory.

DAM FACTS

Height
726.4 feet

Width
1,244 feet
(along the top)

Thickness
45 feet
(at the top)
660 feet
(at the bottom)

Weight
6,600,000 tons

**Most workers on the
job at one time**
5,218

Start date
March 11, 1931

Completion date
February 29, 1936

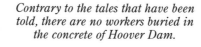

*Contrary to the tales that have been
told, there are no workers buried in
the concrete of Hoover Dam.*